I0481252

Learning Geometry

A Quick Guide to Understanding the Basic Geometrical Concepts

David Itanola

Copyright © 2021 David Itanola

ISBN: 9798713446277

Table of Contents

Introduction

Geometry is everywhere and its application is needed in all aspects of our life. If you are new to geometry or need geometry knowledge to help your child with geometry homework, this is the book for you. Learning Geometry has been carefully written in a simple yet profound manner to help refresh your memory about things you learned while in school. It also teaches you geometry basics, if you are new to geometry. Also, the book is a quick read for anyone preparing for geometry exams. It brings into memory quickly all the Geometrical basics needed for advanced study of Geometry.

1 Understanding Geometrical Concepts

Geometry has many practical applications in professional and real-life situations. Many types of fields, and professions today including engineers, architects, artists medical professionals, real-estate agents, farmers and construction workers require the working knowledge of geometry to complete their day to day task

Geometry is the branch of mathematics that deals with the study of lines, points, angles, surfaces, and solids.

The fundamental geometrical concepts depend on Points, Lines and Planes. They underpin almost every other concept in geometry. You really need to understand these basic concepts, before you can study geometry effectively. They are crucial components that must be well understood in the study of geometry. The idea of dimensions helps us to understand this all important aspect clearly.

Point: A point is usually represented by a dot; it is a single location in space. It has no dimensions as you cannot describe it in terms of length, width or height, and is usually named with an upper case letter.

However, you can use coordinate to describe a point, to indicate its position in space, in relation to a reference point of known co-ordinates.

Nearly everything starts with a point in geometry, whether it's a line, or a three-dimensional object.

• D

Point D

Line: A line is a collection of points that extend endlessly. It is one-dimensional, as it has length, but no width.

The following is a line.

Line segment: A segment has two endpoints. i.e. the starting point and end point. We can simply say; `It is part of a line with starting point and the end point.

X   Y

A line segment XY

Ray: A ray is a type of line that begins at one point and extends indefinitely on one direction.

They are often represented as a line starting from a point with an arrow on the other end: The following is a ray.

Parallel lines. Two distinct lines that never meet in space or on a plane no matter how long they are extended are parallel lines.

Intersecting lines: When lines meet either in space or on a plane, then we say that they are intersecting lines

Perpendicular lines. These are two lines that intersect at a right angle, 90°:

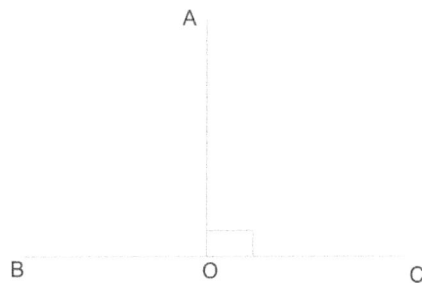

AO is perpendicular to BC

Vertex: vertex is a point where two rays meet. The plural **form is vertices.**

Plane: A plane is a flat surface like the floor. It is also known as two dimensional and extends in all directions.

Two-dimensional figures include polygons such as squares, rectangles and triangles, which have straight lines and a point at each corner.

Three-dimensional shapes include cubes, spheres, pyramids and cylinders. They have height, length and width.

Degrees represented as ° are a measure of rotation, and define the size of the angle between two sides.

2 Introduction to Angles

The study of Angles is the next important components after learning the idea of points, lines and planes

An Angle is formed between two rays that extend from a single point. Or we can that an angle is two rays with the same endpoint.

Angles are commonly indicated using a segment of a circle (an arc). Right angles are 'usually squared off'.

See the diagram below

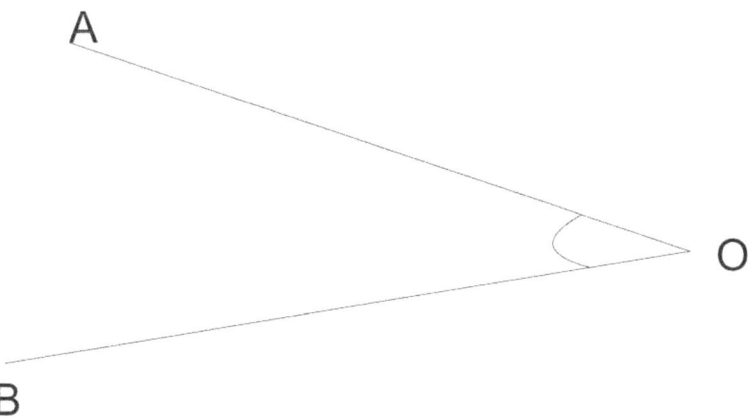

The symbol '∠' is used to describe angles

The expression ∠AOB represent the angle between lines AO and OB. The middle letter in this expression is always the vertex of the angle you are describing.

Furthermore, angles are used mainly to describe shapes such as polygons and polyhedrons, and to explain the behavior of lines.

It's crucial to become familiar with some of the common terminology, and know how to measure and describe angles.

Properties of Angles

Angles are measured in degrees and radians; a full rotation is 360°. A half-circle is 180°, and a quarter-circle, or right angle, is 90°.

Types of angles

There are various types of angles

Acute Angle

An angle that is less than 90 degrees is acute

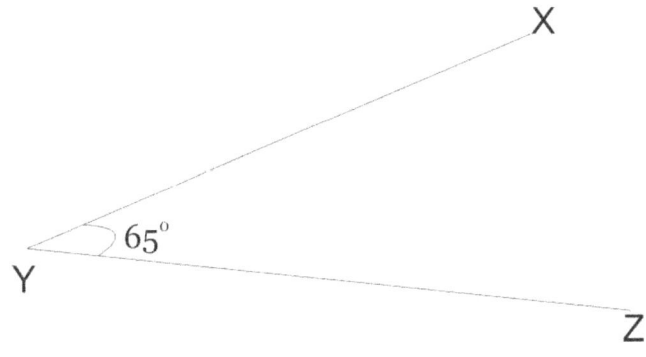

Obtuse Angle

An angle that is more than 90 degrees but less than 180 degrees is known as obtuse. It lies between 90 degrees and 180 degrees.

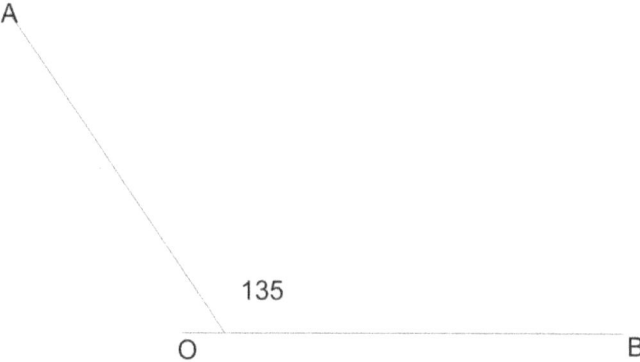

Right angle

An angle that measure 90 degrees is right angle. The following is a right angle.

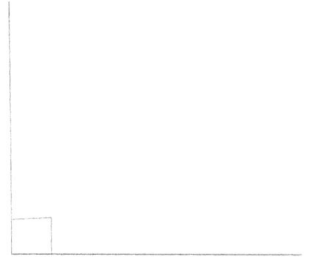

An angle of exactly 180° is said to be angle on a straight line. Thus, a straight line angle looks like a straight line. The following is a straight angle.

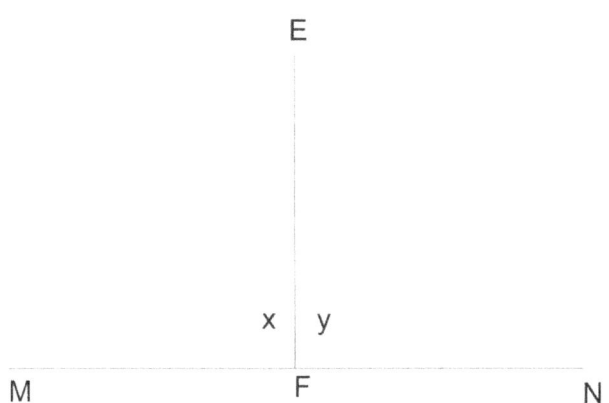

MFN is a straight line angle.

Reflex Angle

Angles greater than 180° but less than 360 degrees are called reflex angles.

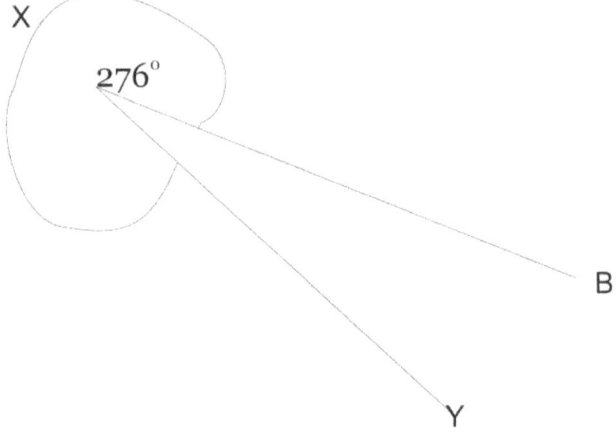

Complementary Angles

Two angles adding up to give 90 degrees are referred to as complementary angles. We can say that the two angles complement each other. For example, if two angles are complementary and one is a the other one is 90-a

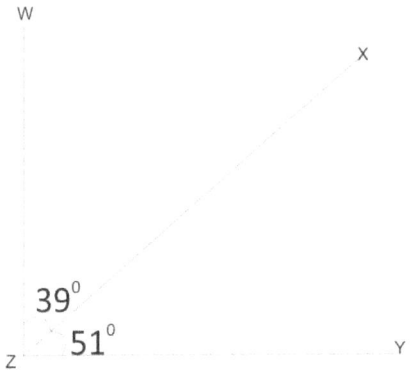

In the above figure, the measure of angle WZX is 39 and angle XZY measures 51.

Adding the two angles we get a right angle, therefore ∠WZX and ∠XZY are complementary angles.

Supplementary Angles

Two angles with sum of 180 degrees are called supplementary angles. If the two angles form a linear angle, such that, if one angle is y, then the other the angle is 180 − y.

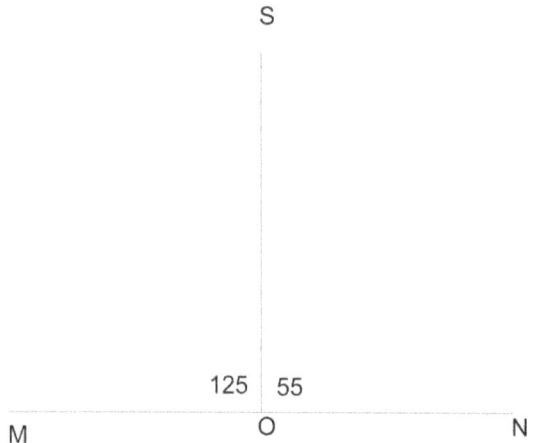

In the above figure, the measure of ∠ MOS is 125 and ∠ SON measures 55. On adding both of these angles we get a straight line angle. Therefore, the pair of angles 125 and 55 are supplementary angles.

Vertical Opposite Angles

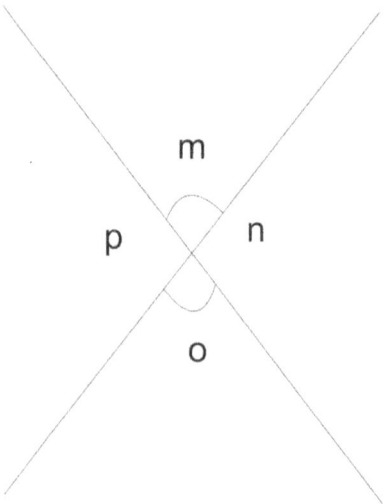

The pairs of m and o and p and n are vertically opposite angles.

When two lines intersect at a point, angles are formed, the formed opposite angles are equal.

If two parallel lines (A and B) are crossed by a third line C(Transversal), 8 angles are formed. then the angle at which the transversal crosses will be the same for both parallel lines. See the following diagram

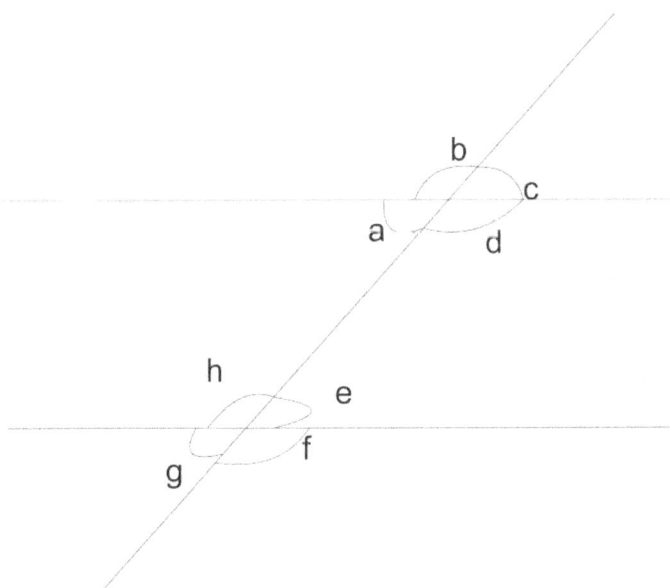

Pairs of vertically opposite angle

a and c

b and d

h and f

e and g

corresponding angles

h and b

g and a

e and c

f and d

alternate angles

a and e

h and d

Measuring Angles

We use protractor to measure angles. They are usually circular or semi-circular and made of transparent plastic, so you can easily see the angle you want to measure

Perpendicular bisector of a line segment

A perpendicular bisector usually written as ⊥ of a segment is a line, segment or ray that is ⊥ to the segment at its midpoint.

Therefore, a perpendicular bisector will bisect a segment into two congruent segments. It can as well be described as the locus of points equidistant from two given points. In the figure below

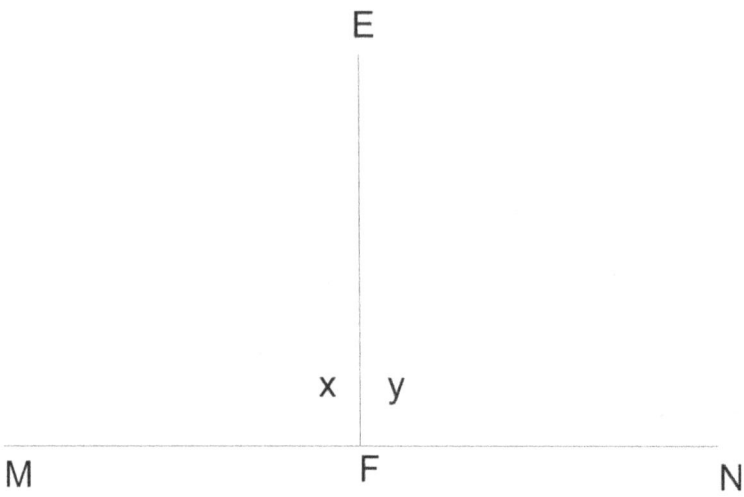

line EF is the ⊥ bisector of segment MN. Angle x and y 2 are right angles.

Perpendicular bisector of a triangle

The perpendicular bisector of a triangle does not necessarily bisects the triangle into two congruent figures but the sides. It is a side of a triangle that it bisects and is perpendicular to a side of the triangle.

Angle bisector

An angle bisector is a a line, segment or ray that divides a given angle into two equal angles or congruent angles

Learning Geometry

3 Introduction to Polygons

A polygon is any 2-dimensional figure (flat shape) formed with straight non-intersecting lines segment. It includes any shape made up of straight lines that can be drawn on a flat surface, like a piece of paper. Polygons are usually defined by the number of sides they contain.

For example, a triangle has 3 sides, and a quadrilateral has 4 sides. In other words, any closed shape that can be drawn by connecting three straight non intersecting lines is called a triangle. So also any shape that can be drawn by connecting four straight non-intersecting lines is called a quadrilateral.

Properties of Polygons

When working with properties of polygons the major areas of interest include:

- The length and **number of sides**
- The **angles** between the sides

When polygons are classified using enclosed angles, then, we have the following;

Regular Polygon: This is a type of Polygon in which all its sides and all the interior angle are equal.

Irregular Polygon: This is the type of polygon which is not regular. It has no equal sides and measure of angles.

Convex Polygon: This is a type of polygon whose measure of an interior angle is less than 180°. Their vertices are always outwards.

Concave Polygons: A concave polygon has at least one of its angles measuring more than 180°. Also, all the vertices of a concave polygon are both inwards and outwards.

Quadrilateral Polygons: These are the polygons which are four-sided or with four angles. Every polygon which forms four angles is known as quadrilateral polygons.

Polygons With Three Sides- Triangles

Types of triangles

Equilateral – Equilateral triangle has all its three sides equal , and all its internal angles to be 60°.

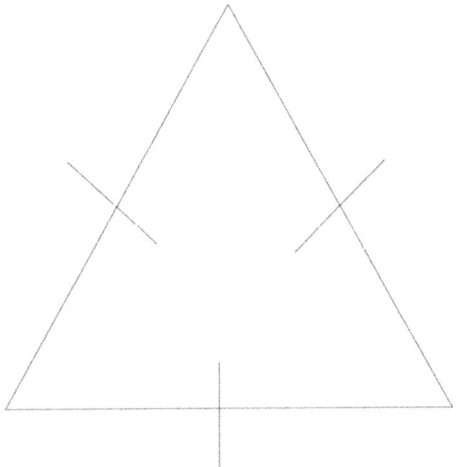

Isosceles – Isosceles triangle has two of its sides equal, with the third one a different length. Its base angles are equal.

23

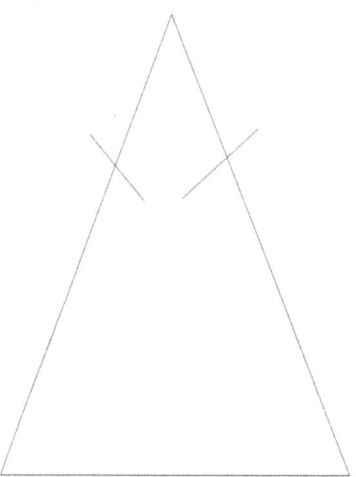

Scalene – This is a type of triangle in which all three sides, and all three internal angles, are different.

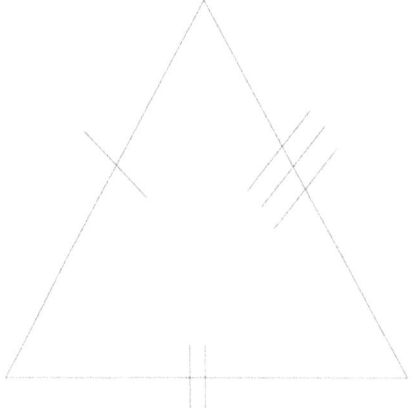

Furthermore, we can describe triangles in terms of their interior angles. The interior angles of a triangle always add up to 180°. In this regard, a triangle with only **acute** interior angles is called an acute-angled triangle.

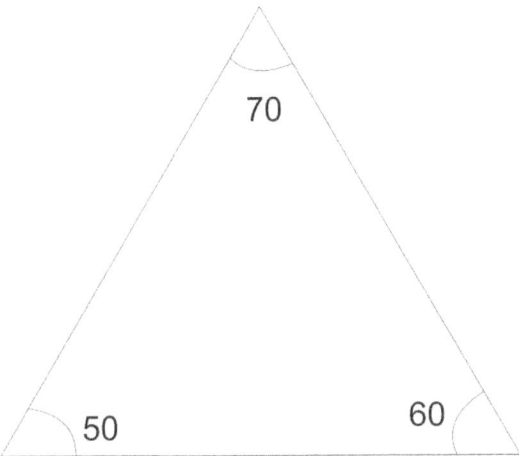

Note that a triangle cannot have more than one obtuse angle. Therefore a triangle having one of its angle obtuse and two acute angles is called obtuse – angled.

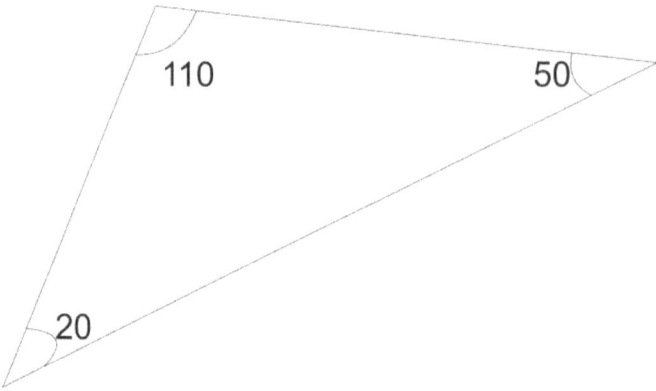

Also, a triangle having one right angle is called a right-angled.

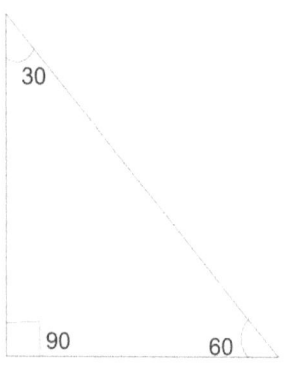

Each of these further classifications will *also* fall under the general triangle classification as they will either be equilateral, **isosceles** or **scalene**.

Four-Sided Polygons - Quadrilaterals

Four-sided polygons are known as quadrilaterals. They are sometimes known as quadrangles or tetragons.

The interior angles of a quadrilateral add up to 360°.

A type of quadrilaterals includes the square, rectangle, kite and other parallelograms, trapezium/trapezoid and rhombus.

Square: a square has four equal sides with all its angles right angles.

Rectangle: a rectangle has opposite sides of equal length with four right angles.

Parallelogram: Opposite sides of a parallelogram are equal and parallel. Also, opposite angles are equal.

Trapezium (or trapezoid): Two opposite sides of a trapezium are parallel, but the other two sides are not. The length of sides and angles are not the same.

Kite: Two pairs of adjacent sides of a kite are of equal length; it has an axis of symmetry.

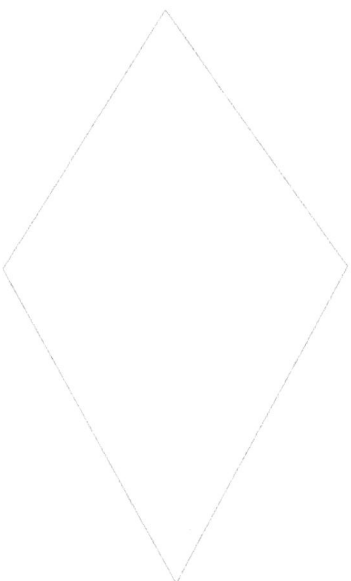

Sum Of Angles of A Polygon

Interior Angle of Polygons

The interior angle of a polygon is the angle formed inside a polygon that lies between two sides of a polygon.

The sum of angles of a polygon is the total sum of all its interior angles. The formula for finding the sum of angles of a regular polygon is given by;

Sum of interior angles = $(n - 2)180$ or $(2n-4)$ right angles

Where n stands for the number of sides of a polygon.

Examples

1. Sum of Angles of a Triangle:

The number of sides of a triangle is 3, therefore,

$n = 3$

Using 3 for n in the formula we have

Sum of interior angles = $(n - 2)180$

$(3 - 2)$x $180=$

1x$180=$

180°

Using the second formula we

$\{(2\text{x}3)\text{-}4\}$x $90=$

$6\text{-}4$x $90=$

2x$90=$ 180

The sum of angles of a triangle is 180

2 Sum of interior Angles of a Quadrilateral:

A quadrilateral has 4 sides, therefore,

$n = 4.$

Substituting this into

(n – 2)180 gives

2x180

$=360°$

The sum of angles of a quadrilateral is $360°$

3. Sum of interior Angles of a Pentagon

A pentagon has 5 sides therefore

n = 5

Substitute 2 for n we have.

(5 – 2) x180

$= 540°$

4. Sum of Angles of a nonagon.

A nonagon is a 9 – sided polygon, therefore

n = 9

When you substitute 9 for n in our formula we have

$(9 - 2)$ x180

=7x180

=1260^0

Learning Geometry

4 Calculating Area of Plane Shapes

Calculating the area of a shape or surface is an important aspect of geometry which can be very useful in everyday life. For example, if you need to know how much carpet that is needed to cover a floor, the knowledge helps you to buy the exact amount that is needed. This chapter covers all you need to know to understand and calculate the areas of common plane shapes (squares and rectangles, triangles and circles).

Calculating Area Using the Grid Method

This is a simple method for calculating areas of plane shapes. It involves calculating area by counting the number of grid squares inside the shape.

it works well for all shapes – as long as the grid sizes are known. The only problem is when the given shapes do not fit the grid exactly, or when you need to count fractions of grid squares. In that case you have to be more discreet to calculate the area.

Rectangle and Square

To find the area of a rectangle or square just multiply its height by its width. Remember that a square has the same length and breadth. All you need to do is to square the side you are given.

From the diagrams below; by counting the square units in the enclosed area, we can see that the first diagram contains 40 square units. This means that the area or the space covers by a rectangle of sides 8 by 5 units is 40 square units. Similarly, the second diagram shows that a rectangle of 10 by 7 units covers an area of 70 square units.

Calculating Area of Plane Shapes

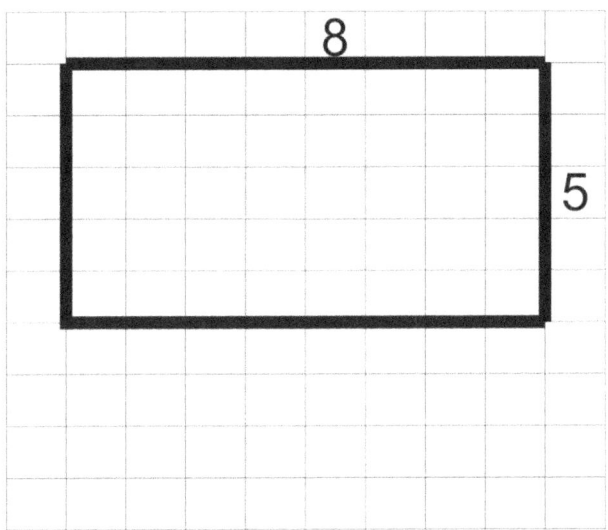

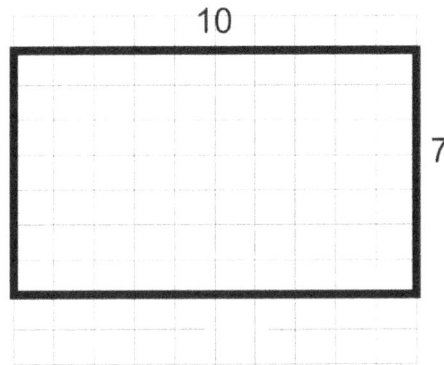

Note that 8x5=40 and 10x7=70

Thus, area of a rectangle can be obtained by simply multiply its length by the breath. So a rectangle that has a length of 8units and breadth of 7 units will enclose an area of 56 square units.

Examples;

1. Calculate the area of a rectangle whose length is 18cm and width15cm.

Solution

Area= length x breadth

   (18 x 15) cm²

   = 270 cm²

2. If the area of rectangle floor is 6450cm and its length is 86cm. find the breath

Solution

Area of the floor = 6450cm²

Length of the floor =86cm

Breadth= 6450/86 = 75cm

Breadth= 75cm

3. The area of a square is 1440m². Find the length of its sides.

Solution

The area of a square is 1440m².

A square has 2 equal sides

Its side= square root 1440 = 120m

4. In a kitchen of sides 45m by 9m, an area of 50m² is reserved for cabinet. What area is left for cooking space?

Solution

The area of the kitchen (45 x 9)m = 405m²

Area marked for cabinet is 50m²

Therefore, the remaining kitchen area = 405-50=355m²

5. How many tiles of dimension 40cm by 40cm are needed to cover a rectangular floor of length 12.5m by 7m.

Solution

Area of 1 tile = (40x40)cm = 1600cm²

Area of the floor (12.5 x 7)m in cm

(1250x700)cm² = 875000cm

Since1 tile covers 1600cm; therefore, the tiles needed for the rectangular floor = 875000/1600

= 546.875 tiles

Area of parallelogram

Consider the parallelogram; WXYZ of height h

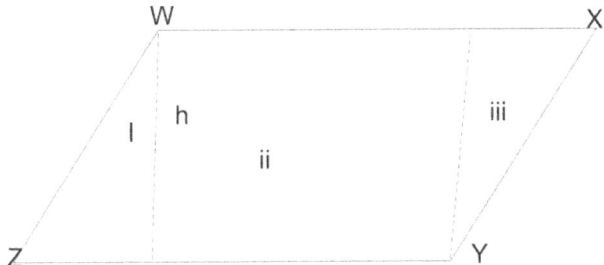

If we cut off triangle I in the parallelogram WXYZ and re-arrange the order by moving the cut triangle to the side of triangle III. It will give something like this

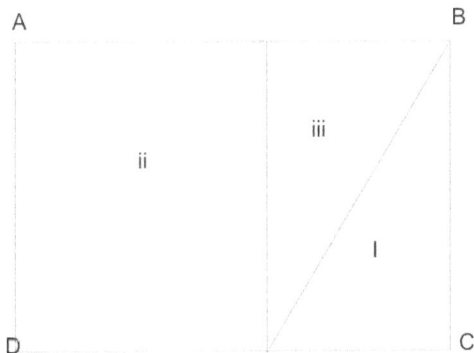

Which is a rectangle. Let us name it ABCD.

The most interesting fact is that the area of the parallelogram WXYZ is the same as the area of rectangle ABCD.

Note;

i.   h is the height of both the parallelogram WXYZ and the rectangle ABCD

ii.   b is the base of rectangle ABCD which is also the base parallelogram WXYZ and DC= AB=ZY=WX=b

Recall that the area of a rectangle is length x breadth i.e b x h in this case

Area of parallelogram WXYZ, is the same as rectangle ABCD

Hence, for any parallelogram with base b and height h its

Area= b x h

Similarly, we can equally deduce this formula if we consider the area of ABCD as sum of two equal distinct triangles that made up the parallelogram as it can be seen below. AC bisects parallelogram ABCD into two equal triangles.

Calculating Area of Plane Shapes

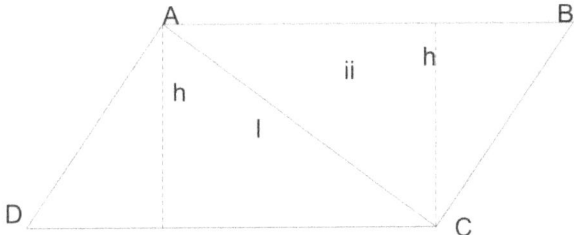

The area of triangle I ADC in 1/2 bxh= bh.

The area of triangle II ACB is ½ bh

Area of parallelogram

ABCD= 2(½ bxh)= bh

Examples

a. Calculate the area of a parallelogram whose height is 9m and base 6m.

solution

base=6

height = 9

area = b x h ; 9 x 6= 54m²

b. calculate the height of a parallelogram whose area = 35m² and base equal 7/2 m.

Solution

Area = 35m²

Base= 7/2 m

Height= area/base = 35/7/2 = 35/1 x 2/7 = 10m

Exercises

1 Calculate the area of the following parallelograms below;

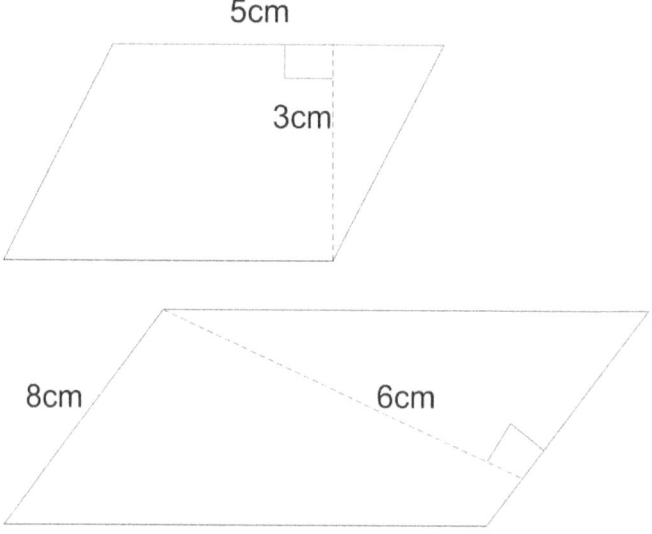

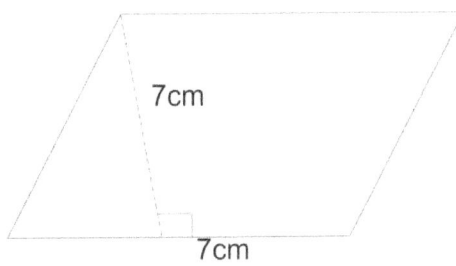

2. The area of parallelogram is 48cm². If the length of the base is 12cm, what is the height of the parallelogram?

Area of Triangle

Consider the figure below

Rectangle ABCD

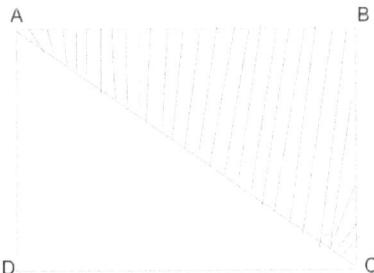

With diagonal AC bisecting the rectangle into 2 equal right angles.

If the length is 8cm and breadth is 6cm. therefore, the area of rectangle ABCD is 8 x 6= 48cm²

The area of triangle ABC is half of the area of the rectangle. i.e ½ (8 x 6) cm²=24cm²

But the base of the triangle is 8cm and its height is still 6cm. the same with the base and height of rectangle ABCD. We can therefore say that the area of a triangle is (½ base x height) square unit. Thus, we can easily find the area of a triangle, if its base and height are known.

Note that the height is always at perpendicular to the base but does not necessarily divide the base in to two paths. It can even lie outside the given triangle.

Example

Find either length or height or area of the figures below

    a.

Calculating Area of Plane Shapes

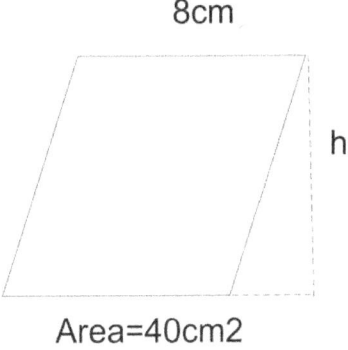

8cm

h

Area=40cm2

b.

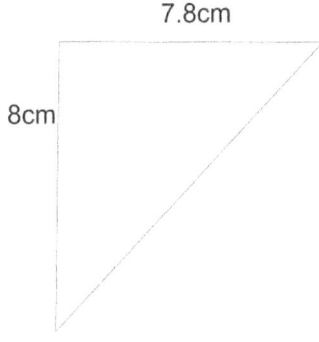

7.8cm

8cm

c

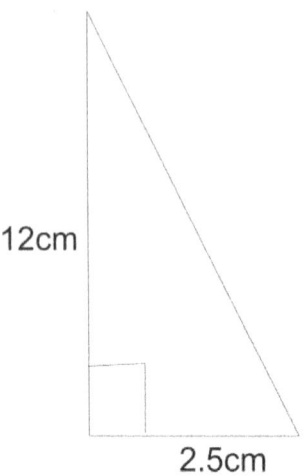

12cm

2.5cm

Solution

a

Area- 40 cm²

Base= 8cm

area – ½ x b x h

40= ½ x 8 x h

h= 40/4

=1 0cm

b

Base=7.8cm

Height= 8cm

Area= ½ x 7.8 x 8

 31.2cm²

c

Base= 2.5cm

Height= 12cm

Area= ½ x 2.5 x 12

15cm²

Exercises

1. calculate the area of the figures below

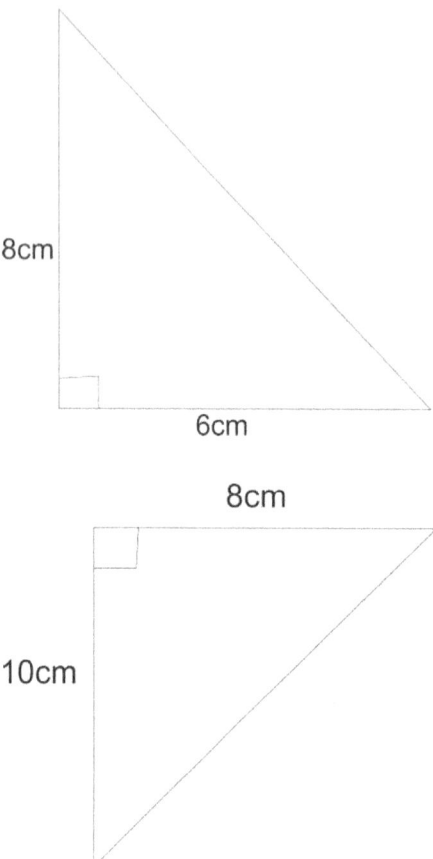

8cm

6cm

8cm

10cm

2. find the length of a triangle whose area is 30cm² and base of 16cm.

3. Calculate the area of a triangle whose height is 5 ½ cm and base is 7cm.

4. Find the length, if the area of a triangle is 38cm2 and the base is 14.5 cm.

Area of Trapezium

Consider the figure bellow

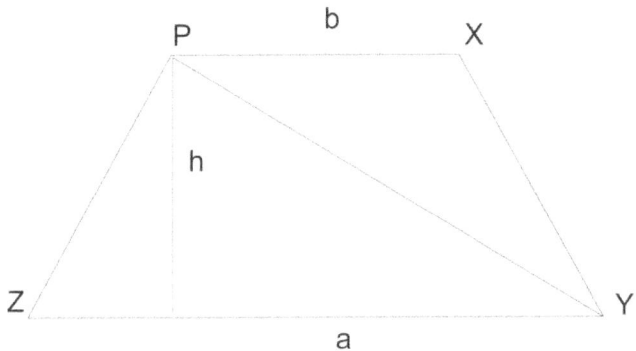

PXYZ is a trapezium with PX // ZY and h being the perpendicular distance between the two parallel sides.

The area of the trapezium is the sum of the areas of the two triangles PYZ and PXY

Area of PYZ=1/2 (a x h)

Area of PXY=1/2 (b x h)

Thus are of trapezium PXYZ= ½ (a x h) + ½ (b x h)

Learning Geometry

$\frac{1}{2} (a \times h) + \frac{1}{2} bh =$

$\frac{1}{2} (a+b)h$

In other words

Area of trapezium= $\frac{1}{2}$ (sum of parallel sides) x the distance between them

Worked examples

1. Calculate the arc of the trapezium WXYZ below

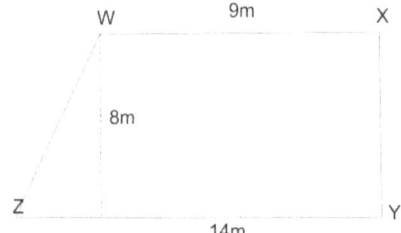

Solution

Using the formula

Area=$1/2 (a+b)h$

a=14m   b=9m   and   h=8m

$\frac{1}{2} (14+9) \times 8 = 23 \times 8 = 92m^2$

2. A plot of land has the shape of a trapezium. If the parallel sides are respectively 120m by 80m and the distance between the parallel sides is 60m, calculate the area of the plot of land.

Solution

$\frac{1}{2}(a+b)$ h m²

a=120   b=80   and h=60m

$\frac{1}{2}$ x (120+80) x 60

=200 x 30   = 6000m²

3. The longer of the parallel sides of a trapezium is 14m. If the distance between the parallel sides is 20m and the area of the figure is 250m², find the shorter side.

Solution

A= $\frac{1}{2}$ (a+b) h

Shorter side=?

Longer side= 14m

Area= 250

Height= 20m

½ (a+14)x 20 = 250

(a+14) x 10 =250

10a + 140 =250

10a = 250-140

10a = 250 − 140

10a=110

a=11m

Exercises

1. Calculate the areas of the diagrams below;

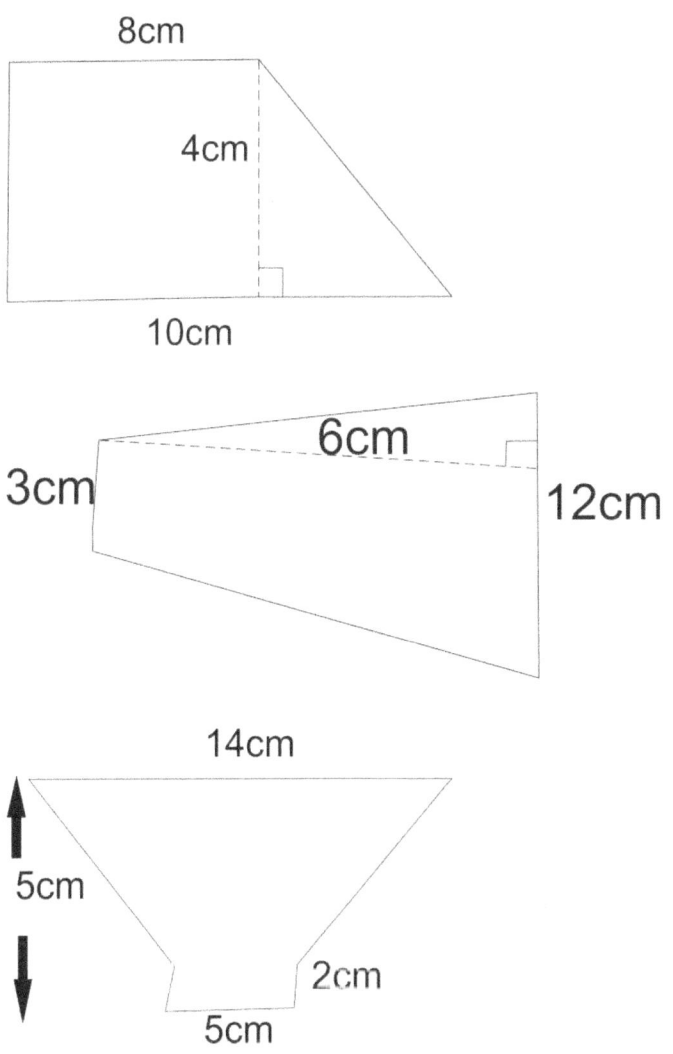

2. The length of two parallel sides of a trapezium , are 5cm and 10cm respectively. If the area of the trapezium is 30cm². What is the

perpendicular distance between the two parallel sides?

3. Calculate the area of trapezium PQRS below;

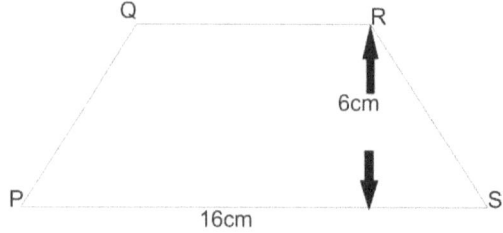

4. Given that A = ½ (a+b)h, if a=6cm and h=3cm, find b when area is 48cm².

Area of sector and segment of a circle

An arc

An arc of a circle is a part of the circumference/perimeter. A sector is the portion of a circle enclosed by two radii. A segment is the part cut off from the circle by a chord.

Area of Circle

A circle has a radius which is a constant distance between the center and the circumference.

Area of a circle is πr^2

π is a Greek word that takes a constant value of 22/7 or 3.142.

Worked examples

1. Find the area of a circle of radius 2.8cm

Solution

Area=πr^2

if r-2.8

then

a= 22/7 x 2.8 x 2.8

=24.64cm²

2. Find the area of a circle whose radius is 31/2cm.

Solution

$a=\pi r^2$

r= 31/2 = 7/2

πr^2

22/7 x 7/2 x 7/2 = 49/4

Therefore; a=36.5cm²

Find the area of a semicircle with a diameter of 28cm.

Solution

If diameter is 28; r=28/2 = 14

Area of semi-circle = ½ πr^2

½ x 22/7 x 14 x 14 = 308cm²

3. The area of sector of a circle is 44cm² . what is the radius of the circle if the angle at the center of the circle is 140⁰?

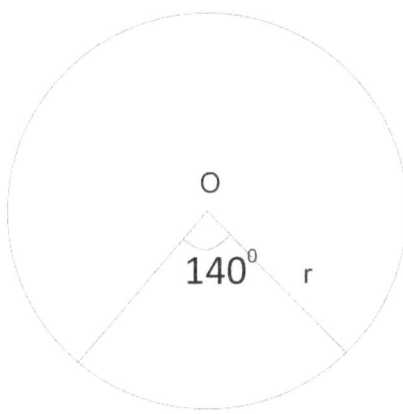

Solution

Area= 140/360 x 22/7 x r² cm²

44=11 x r²/9

11 x r² = 44 x 9

= 44 x 9/ 11

r²= 36

r= 6cm

Exercises

1. A circular drum has a diameter of 40m. Find its radius and hence calculate the area of the drum. Use the value 3.14 for π

2. What is the area of circle of diameter 28cm. take 22/7 for π

3. Find the area of the shaded portion in the quadrant of a circle below

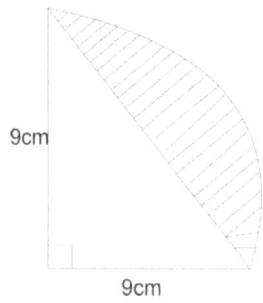

9cm

9cm

4. Find the radius of a circle whose perimeter is 44cm. take $\pi = 22/7$

5. What will be the area to carpet a circular room of radius r?

6. Find the perimeter of a circle, whose diameter is 16cm

5 How to Find the Area of a Border – a Shape within another Shape

There are several ways to work out the area of a shape within another shape. You could break the shape into different regular shapes and find their dimensions and then their area and finally add the areas together to get the total area.

A faster approach would be to work out the area of the whole shape (bigger shape) and the area of the internal shape (smaller shape). when you subtract the area of the smaller shape from the bigger shape what you have left is the area of the border.

Worked example

Find the area of the shaded portion in the diagram below

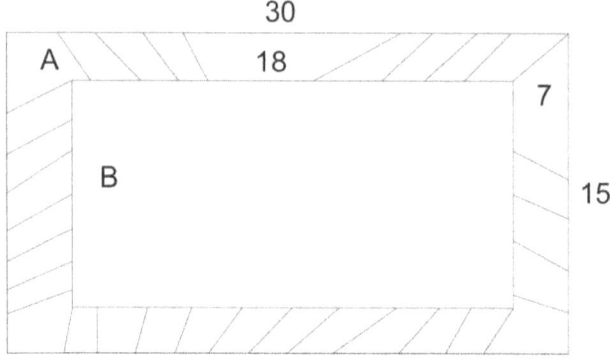

Area of the rectangle A = 30 x 15 = 450cm²

Area of the rectangle B = 18 x 7 = 126cm²

The area of the shaded portion is 450-126 = 324cm²

Exercises

Calculate the areas of shaded portion of the diagram below

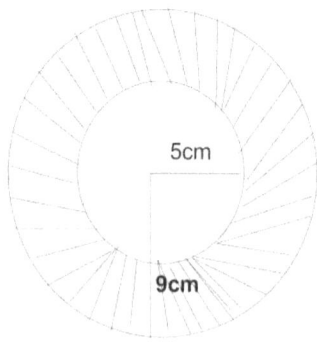

How to Find the Area of a Border

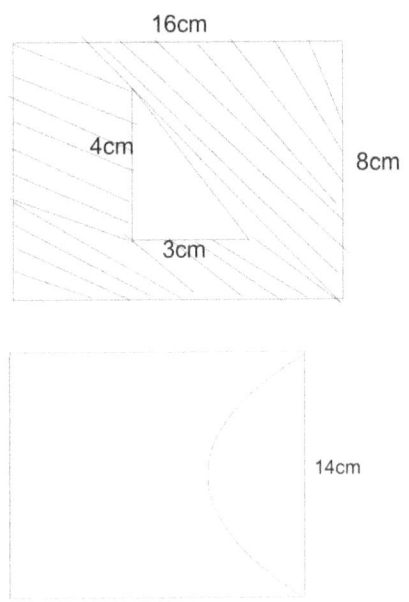

16cm

4cm

8cm

3cm

14cm

18cm

Learning Geometry

6 Area of Irregular Shapes

To calculate the area of irregular shapes, first split the irregular shape into several regular shapes that you can recognize such as triangles, rectangles, circles, squares as the case may be . This can be done in no particular order. Calculate the area of each regular shape and add to get the area of the irregular shape. the following examples give a better illustration.

Examples

Calculate the areas of the shapes belowa

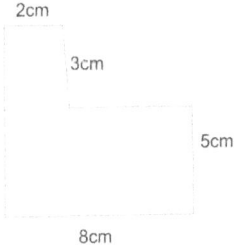

2cm

3cm

5cm

8cm

b

Learning Geometry

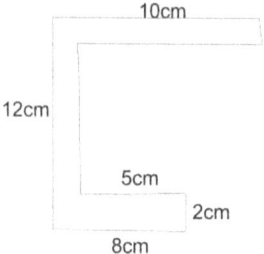

c

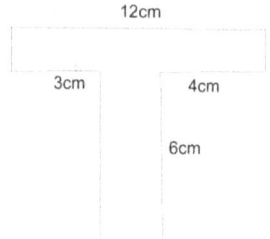

Solution

a.

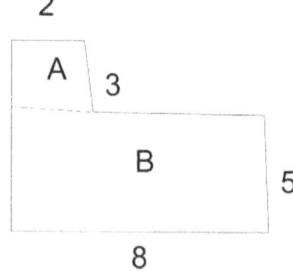

The figure above can be divided into two regular shapes. Rectangles

Area of rectangle A= 2 x 3=6m²

Area of rectangle B = 8 x 5=40m²

Total area= 40m²+ 6m²= 46m²

b.

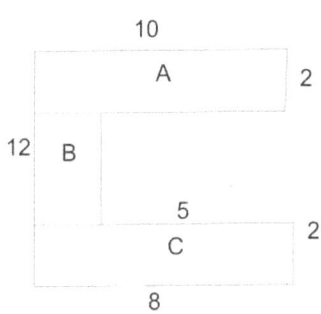

The figure can be divided into 3 rectangles. A,B ,C

Area of A= 10 x 2 =20

Area of B

Length of B= (12- 2+2) =8 and the width = 5.

Area = 8 x 5 = 40

Area C= 2 x 8=16

Therefore, total area = 20+40+16=76cm

c.

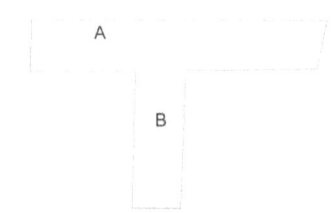

Area of A= 12 x 2=24

Area of B= 6 x 5=30

Total area = 24 + 30=54m²

Exercises

Calculate the areas of the following shapes.

Area of Irregular Shapes

1

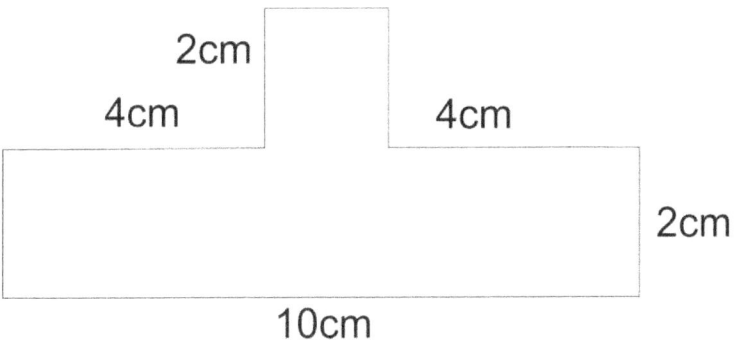

2cm
4cm   4cm
2cm
10cm

2

5cm
5cm
2cm
10cm

3

Learning Geometry

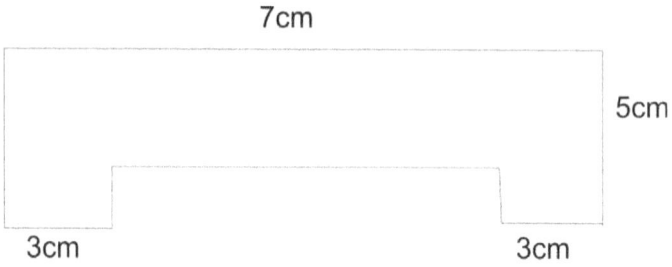

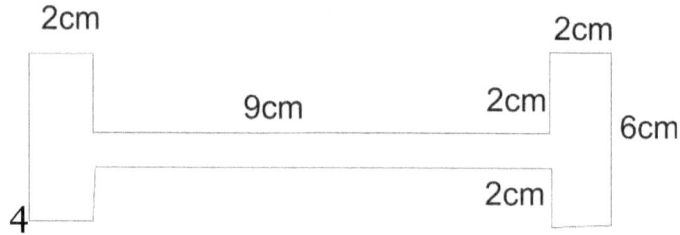

7 Perimeter of Plane Shapes

Perimeter is the measure of the distance round the boundary or edge of any shape.

For any plane shape, its perimeter is the total distance along the outside of the shape.

Calculating Perimeter has many practical applications in everyday life. For example, it could be used to find out the distance ran by a runner around a field, the revolutions of a wheel, length of a fence surrounding a garden etc.

Perimeter of some regular polygons

Regular shapes have their edges make up of smooth curve or straight lines.

 You can measure the edges with the aid of a ruler and add all the sides. The other method is to use the formulae.

Square and Rectangles

The perimeter of a square with sides shown below is a+a+a+a = 4a.

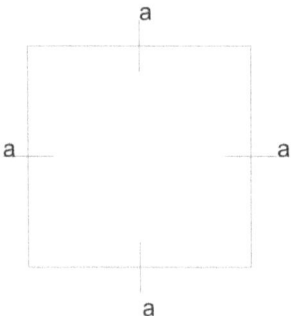

The perimeter of square is four times the height of one of its equal sides. The perimeter of a rectangle with sides of length a and breadth b = a+b+a+b = 2a + 2b = 2(a+b)

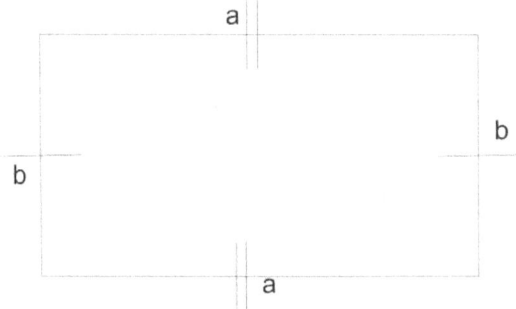

The perimeter of a rectangle is twice the sum of the length and breadth.

Parallelograms

The perimeter of a parallelogram is similar to that of a rectangle. It is given as 2(l+b).

Examples

1. Find in cm the perimeter of the regular pentagon ABCDE with each edge 7cm.

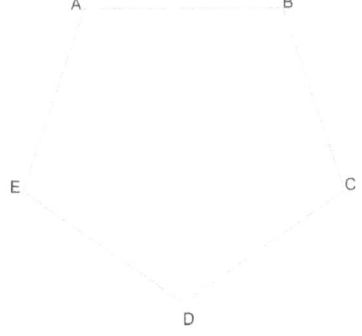

Solution

Length of side AB = 7cm

There are 5 equal sides

Therefore; perimeter = 7 x 5 = 35cm.

OR

AB+BC+CD+DE+EA

= 7cm+7cm+7cm+7cm+7cm\

= 35cm

2. Calculate the perimeter of the shape below

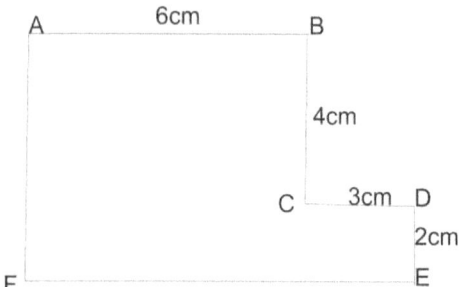

Solution

The perimeter of the shape is AB+BC+CD+DE+EF+AF

AB= 6cm; BC= 4cm; CD= 3cm; DE= 2cm; EF= (6+3)=9cm and AF= 4+2=6cm

Thus the perimeter of the shape = 30cm

3. The perimeter of a rectangular lawn is 50cm. if the length of the lawn is 15cm. what is the width?

Solution

Perimeter of Plane Shapes

Let the breadth be y

$2(15+y)=50$

$15+y=25$

$y=25-15$

$y=10cm$

Exercises

Find the perimeter of the following shapes

1.

2.

Learning Geometry

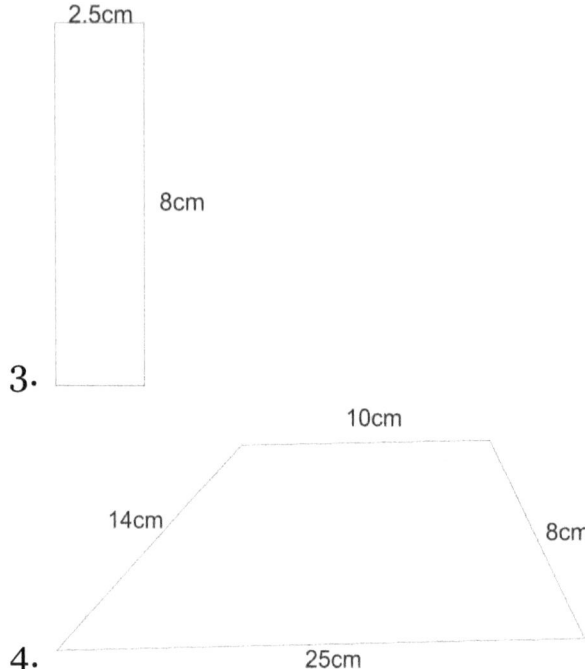

3.

4.

5.

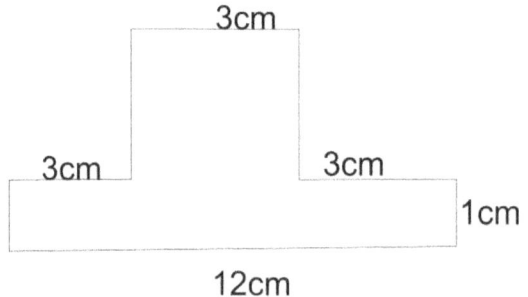

Circles

The perimeter of a circle is also called the circumference of the circle. You can simply obtain the diameter of a circle by winding a thread or string round the circle after which it is rolled and the length of the thread or string is measured with a ruler or calculated by using the formula.

P=2πr   or πd

Where r= radius;   and d= diameter

π= 22/7 or 3.142

Examples

1. Find the circumference of a circle whose diameter is 35cm.

c= πd

c= 22/7 x 35/1

c= 110cm

2. Find the radius and diameter of a circle whose circumference is 220cm.

Solution

$c=\pi d$

c= 220;   d=?

220= 22/7 x d

d= 220 x 7/22

d= 70cm

r= ½ x 60 = 35cm

Therefore; d= 70cm   and   r= 35cm

Exercises

1. Find the radius of the circle with circumference
   i.   115.06m
   ii.   65.36m
2. The diameter of a bicycle wheel is 85cm. what is the distance moved by the wheel in 5 revolutions?

8 Three Dimensional Figures

All objects that have internal and external features are called solid or three-dimensional objects.

Basic properties of some solids

(cuboid, cube, cone and cylinder)

All 3-dimensional objects have surfaces or faces, edges and vertices.

Face ;The outer view of a solid is known as its surface.

Edge; This is the part of a surface where two faces meet. It may be straight or curved.

Vertex; This is a point or corner where 3 or more edges meet.

Properties of a cuboid

Faces; 6

Edges; 12

Vertex; 8

The face of a cuboid could be plane or curved with a rectangular shape. Examples are box, books, and block.

Cube

A cube has all the properties of a cuboid except that all the 6 faces are square. Examples are sugar, dice.

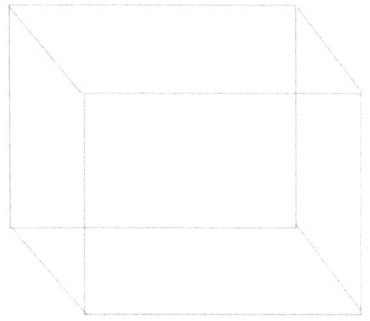

Cone

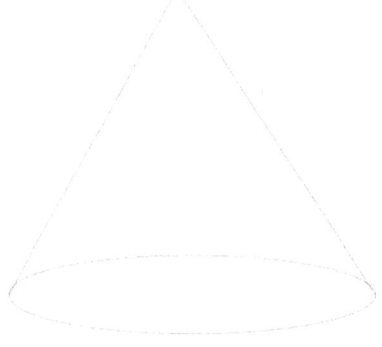

A cone has a circular base and a curved surface. it has only 1 vertex and 1 edge. Examples are roof top, sharpened end of a pencil etc.

Cylinder

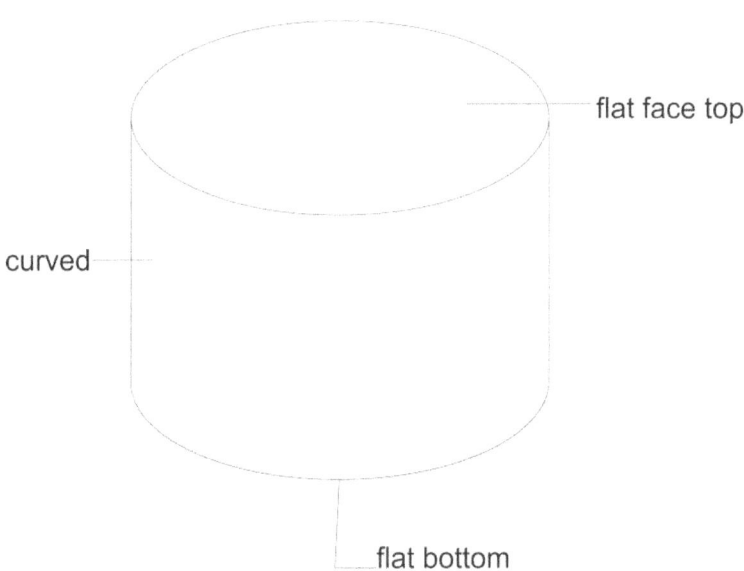

The surface of a solid cylinder consists of three faces. Two flat faces (top and bottom) curved surface.

Prism

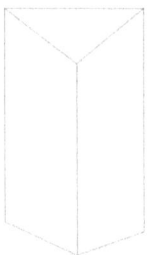

Triangular prism

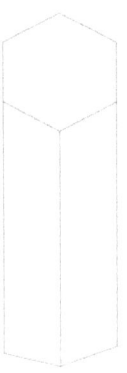

Hexagonal base prism

Prism got its name from the shape of its base and top faces, and its sides are always rectangular. A cuboid can be described as rectangular prism while the cylinder is a special type of prism.

The property of a prism depends on its type. For instance, a triangular prism has 3 rectangular and 2 triangular faces. It has 9 edges and 10 vertices. Hexagonal prism on the other hand has 2 hexagonal faces and 6 rectangular faces with 6 vertices and 9 edges.

Pyramid

There are three types of pyramids each is named according to the shape of its base;

   i.   Triangular based pyramid

   ii.   Square based pyramid

   iii.   Hexagonal based pyramid

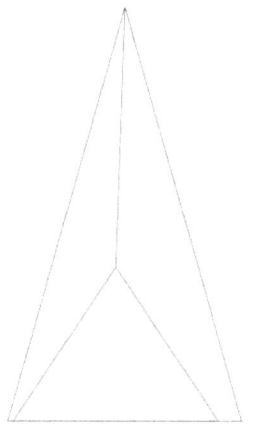
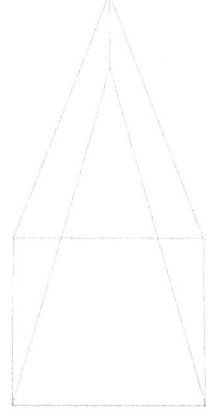

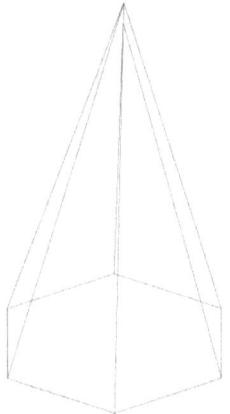

The type of pyramid determines its properties and the shape of its face determines the number of its faces.

The rectangular/square based pyramid has 5 faces, a pyramid that has a base of 5 sides figure has 6 faces. The slant edges also varies with the number of sides it has.

Rectangular/square based pyramid has 4 slant edges and 4 base edges.

Sphere

A sphere is spherical in shape. the shape of a ball, orange, earth and spherical. Half of a sphere is hemisphere.

Unit of measuring volume and capacity

Cubic meter is the standard unit for measuring volume. To measure small volume we use a unit cube of edge/meter.

The volume of a is $1m^2$

$100cm=1m$

$1m^3= (100 \times 100 \times 100)cm^3$

$= 1000000cm^3$

$1m^3=1000000cm^3$

Also, $1cm=10mm$

Therefore, $1cm^3=1000mm$

$1 \text{ liter} = 10 \times 10 \times 10=1000cm^3$

The capacity of a container is the measure of the space occupied by 1 liter at standard temperature and pressure.

Volume of cube and cuboid

The measure of space occupied by a solid shape is the volume of the shape.

Volume= length x breadth x height

Area of base x height

Worked examples

1. Find the volume of a cuboid whose length, breadth and height are 30cm by 15cm by 10cm respectively

V= 1 x b x h cm³

30 x 15 x 10

= 4500cm³

2. The volume of a cuboid is 4275cm³. If its base is 25cm by 18cm. find the height of the cuboid

Solution

V= l x b x h

To find the height, we use the relation;

H= v/l x b

Given; v= 1275cm³

L= 25cm and b=18cm

h= 4275/25 x 18

h= 9.5

3. Find the volume of water in a tank measuring 5m long, 4m wide and 1.5m deep.

Solution

V= l x b x h

(5 x 4 x 1.5)m³

=30 m³

4. A square base tank of sides 3m is 1.8m deep. Calculate the amount of water it contains if it is ¾ full.

Solution

$v = l \times b \times h$

$v = 3 \times 3 \times 1.8$

$v = 16.2m^3$

$1m^3 = 1000l$

$16.2m^3 = 1620l$

Full tank $= 1620l$

Half tank $= \frac{3}{4}$ of 1620

$= 405 \times 3 = 12150l$

Exercises

1. If the volume of a cube is $25cm^3$, what are the dimensions of the cube?
2. A cube sugar is 4cm wide. Calculate the number of cubes in a box $740cm^3$.
3. The area of one side of a cuboid is $27cm^2$. What is the length, if the width is 2.0cm?

4. A rectangular tank that is 20m long and 35m wide contains 1000m³ of water. How deep is the water in the tank?

5. A room that is 10m long and 5m wide contains 98m³ of air. Calculate the height of the room.

Learning Geometry

9 Pythagoras Theorem

Pythagoras Theorem describes the relation between the three sides. **Of a Right-Angled triangle.** It is also sometimes called the Pythagorean Theorem.

The three sides are shown below.

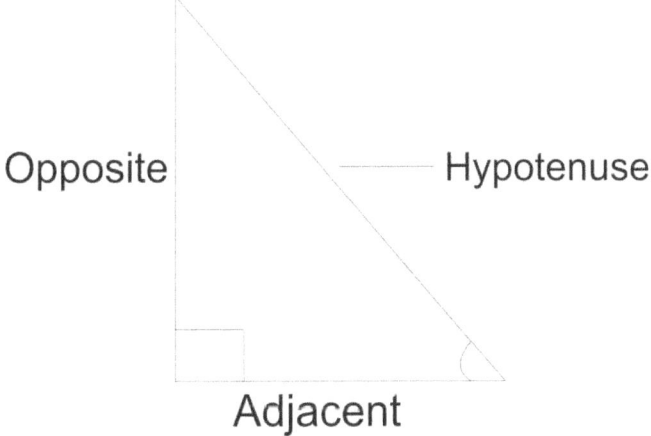

Hypotenuse is the side that faces the right angle and its usually the **longest side** of the triangle . other sides are the base and the height of the triangle. These other sides are not fixed it

depends on how you position the triangle.

Pythagoras theorem states that "In a right-angled triangle, the square of the hypotenuse side equals sum of squares of the other two sides"

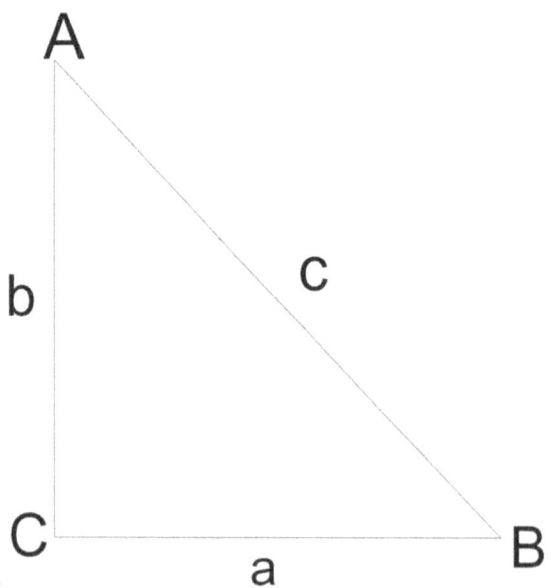

Given a right angle triangle ABC:

Where ""c" is the hypotenuse side,

a" is the height and

"b" is the base,

Then

$c^2=a^2+b^2$

Uses of Pythagoras theorem

The theorem is used to find the length of any side in a right angled triangle given the length of the other two sides.

The theorem helps to know if a given triangle is a right angled triangle or not.

It also helps us to find the diagonal of a rectangle or square

The theorem is mostly used in the field of construction. it can be used to find the steepness of the hills or mountains. .

Examples:

1.Find the value of x. in the right angled triangle, given below:

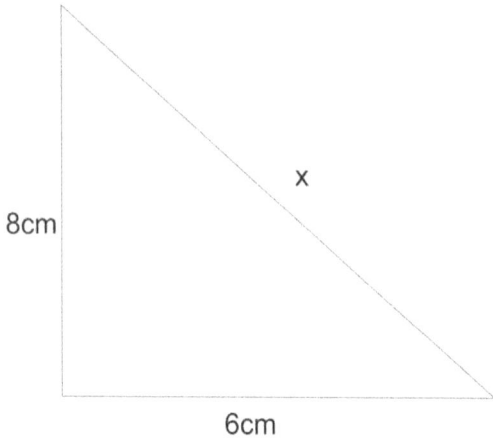

8cm

6cm

This shows that x is the side opposite to the right angle, hence it is the longest side ie the hypotenuse.

Using the theorem

$x^2 = 8^2 + 6^2$

Pythagoras Theorem

$x^2 = 64+36 = 100$

$x = \sqrt{100} = 10$

Therefore the value of x is 10

2. Given the length of sides of a triangle as 5,12 & 13 units. Check if it is a right angled triangle.

Solution: From Pythagoras Theorem, we have;

$a^2 + b^2 = c^2$

Where c is the hypotenuse and a and b the other two sides

Clearly

c = 13 units the longest side

so let

a = 5 units and

b= 12 units

Therefore

$12^2 + 5^2 = 13^2$

$144 + 25 = 169$

95

$169 = 169$

Since

L.H.S. = R.H.S.

Therefore, the triangle is a right angled triangle.

3. Find the value of k in the right angled triangle below .

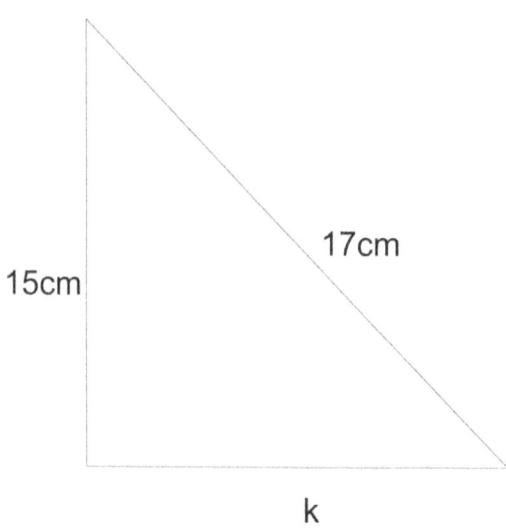

15cm

17cm

k

Solution:

Therefore

$15^2 + k^2 = 17^2$

$225 + b^2 = 289$

$k^2 = 289 - 225$

$k^2 = 64$

$k = \sqrt{64}$

$k = 8$ cm

4. Given a triangle with sides 10, 24, and 26 units. Verify that that triangle is a right angled triangle

Clearly, 26 is the longest side.

Using the Pythagoras theorem

$c^2 = a^2 + b^2$

We can let a = 10, b = 24 and c = 26

We need to show that LHS =RHS to satisfy the condition for Pythagoras theorem.

In the RHS we have

a² + b² = 10² + 24² = 100 + 576 = 676

the LHS gives;

c² = 26² = 676

which shows that

LHS = RHS

Therefore, we can conclude that the given triangle is a right triangle, as it satisfies the condition of the theorem.

5. A ladder 7m long leans against a wall. The ladder's foot is 2m from the wall. Calculate how far up the wall the ladder reaches

Solution

Pythagoras Theorem

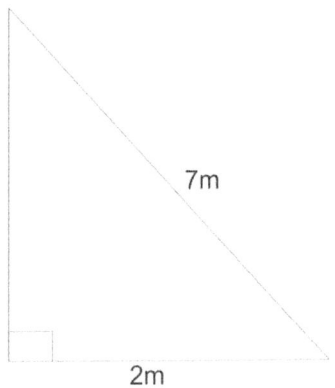

The ladder forms a right angle triangle with the wall. therefore using the Pythagoras theorem we have

$7^2 = 2^2 + y^2$

$49 = 4 + y^2$

$y^2 = 49 - 4$

$y^2 = 45$

$y = \sqrt{45}$

$y = 6.$

6 Find the value of h in the diagram below. It is advisable to first find the value of y^2 before finding the value of n.

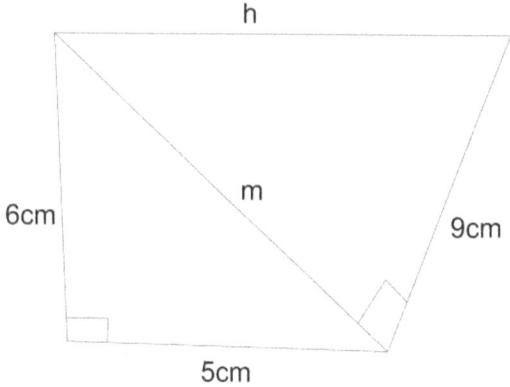

solution

$m^2 = 6^2 + 5^2$

$= 36 + 25$

$= 61$

$h^2 = m^2 + 9^2$

$h^2 = \sqrt{61}^2 + 9^2$

$h^2 = 61 + 81$

$h^2 = 142$

100

$h = \sqrt{142}$

$h = 11.92$

Pythagorean triples

A Pythagorean triple is a set of three whole numbers which give lengths of the sides of right-angled triangles. (5, 12, 13), (7, 24, 25), (8, 15, 17) are some other common Pythagorean triples..

Exercises

1. A ladder 8.5m long leans against a wall. The ladder's foot is 2.5m from the wall. Calculate how far up the wall the ladder reaches

2. Find the perimeter of a rectangle whose length is 60 m and the diagonal is 100 m.

3. The height of two towers is 52 m and 47 m respectively. If the distance between them is 12 m, find the distance between their tops.

10 Similar Triangles

Two triangles are said to be similar if their corresponding angles are congruent and corresponding sides are proportional.

For proper illustration, let us consider triangles ABC and PQR below.

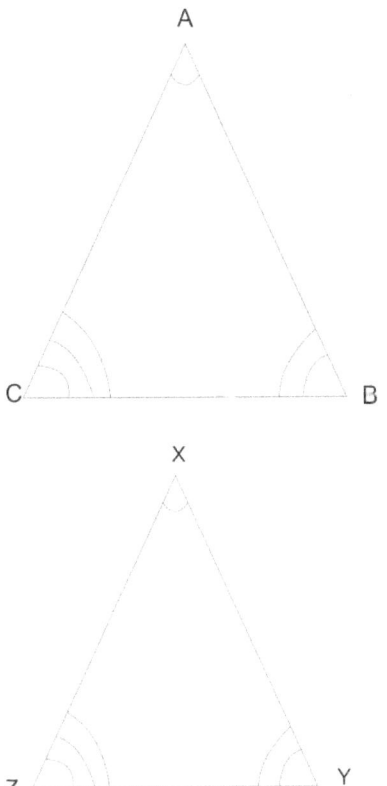

ΔABC and PQR are similar only if,

i)   ∠A = ∠X, ∠B = ∠Y and ∠C = ∠Z

ii)   AB/XY = BC/YZ = AC/XZ

The three triangle similarity theorems that specify under which conditions triangles are similar are discussed below. They are used to check for similar triangles.

These **rules** include: AA rule, SAS rule or SSS rule.

The AA, AAA, angle-angle similarity Theorems.

If two of the angles of two triangles are the same, the triangles are similar.

If two of the angles in a triangle are the same, definitely the third angle will have the same value. This is why this rule is sometimes refer to as AAA.

This becomes obvious in view of the fact that the three angles of a triangle always add up to 180 degrees. If two of the angles are known, the third can be found by subtracting the sum of the two known angles from 180 which gives the value of the third angle.

Furthermore, having two angles equal in a triangle implies that all the three angles are equal. Equilateral triangles are always similar.

SAS or Side-Angle-Side Similarity theorems

Another condition for similarity is SAS. This is a situation in which two of the sides of the two triangles are proportional as well as the included angle.

SSS or Side-Side-Side Similarity theorems

If all the three sides of the two triangles are in the same proportions, then the two triangles are similar.

Solving similar triangle problems

Basically, there are **two types of similar triangle problems**; you may be required to solve. You may need to prove whether a given set of triangles are similar or you can calculate the missing angles and side length .

Similar Triangle Formula helps us to find the dimensions of the other triangle once we know all the dimensions and angles of one triangle.

If triangles ABC and XYZ are two similar triangles, then by the help of similar triangle formulas, we can find the relevant angles and side lengths of triangle XYZ with the details given in triangle ABC and vice versa .

Solved problems

Examples

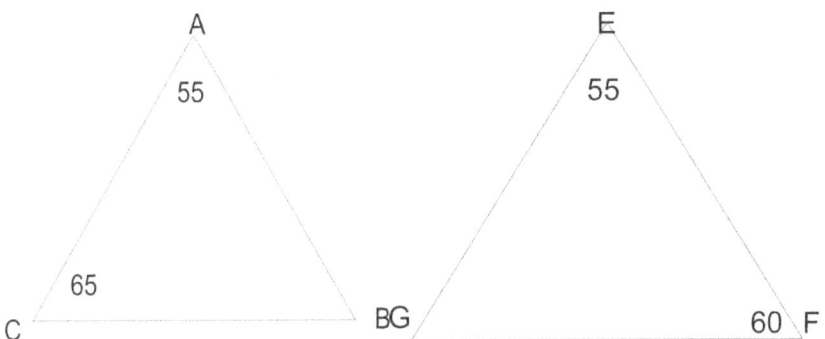

Verify whether triangles ABC and EFG below are similar

Solution

Sum of interior angles in a triangle = 180°

Therefore, by considering Δ ABC

∠A + ∠B + ∠C = 180°

55° + 65° + ∠C = 180°

$120° + \angle C = 180°$

Subtract both sides by 120°.

$\angle C = 60°$

Consider $\triangle$ EFG

$\angle E + \angle F + \angle G = 180°$

$\angle 55° + \angle F + \angle 60° = 180°$

$\angle 115° + \angle F = 180°$

Subtract both sides by 115°

$\angle F = 65°$

- $\angle A = \angle E = 55°$ and $\angle B = \angle F = 65°$

This satisfy Angle-Angle (AA) rule, hence,$\triangle ABC$ is similar to $\triangle EFG$.

Example 2

Given that triangles JKL and MNO below are similar find the value of x.

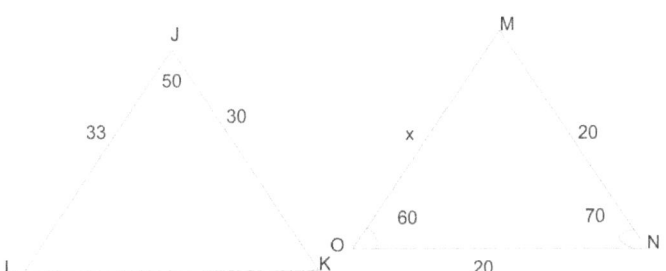

Solution

Given that the two triangles are similar, then;

JK/MN = JL/MO

∠M = 180-130=50

JK/MN=JL/MO

30/20=33/x

30x=660

x= 660/30

X= 22

Example3

In triangle XYZ below

XY= 7cm, MZ= 14cm, and ZY= 28cm

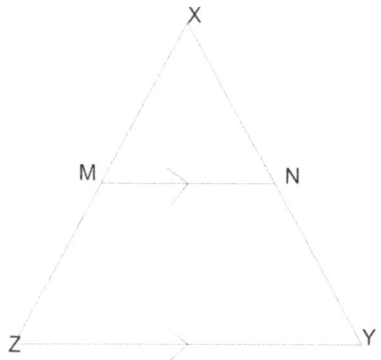

Also, MN//ZY

Find MN

Solution

We need to show that triangles, XMN and XYZ are similar.

∠X is common to both XMN and XYZ. ∠XYZ and ∠XMN are equal; Corresponding angles are equal

Also, ∠XYZ and ∠XNM are equal.

Therefore, XYZ and XMN are similar triangles since they both have two corresponding congruent angles.

Therefore, XM/XZ=MN/ZY

7/21=MN/20

21MN=140

MN= 140/21

MN= 6.6cm

Exercises

1.   Check whether the following triangles are similar

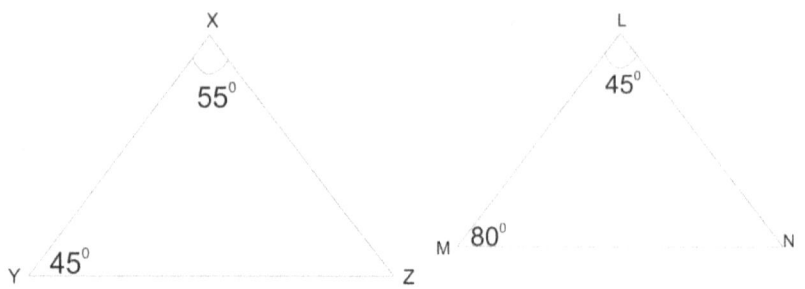

2. Find the value of m in the following triangles if 60⁰

ΔEFG is similar to IJK.

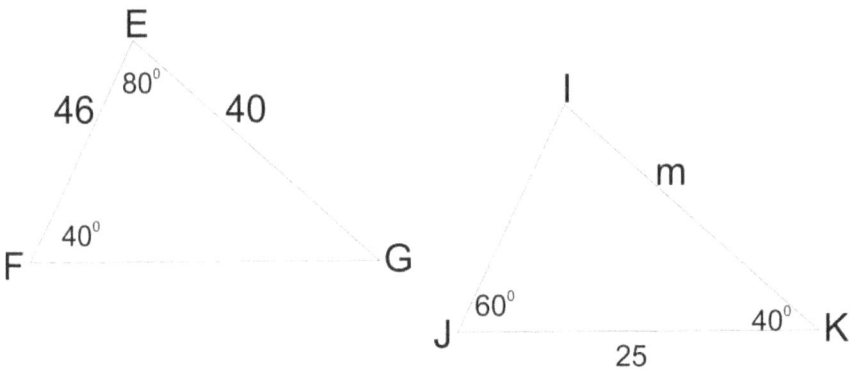

Zx23. Determine the value of k in the following diagram.

Similar Triangles

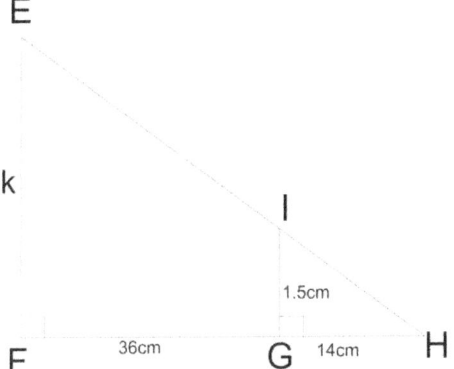

Learning Geometry

About the Author

David Itanola has taught mathematics at various levels for more than 30 years.. Apart from writing on mathematical subjects, David has a penchant for helping people living a productive life.

David holds the master's degree in Mathematics. He is married with three children and enjoys playing the piano and mentoring children.

Learning Geometry

Resources

https://www.radford.edu/~wacase/**Geometry** Notes 1.pdf

www.toppr.comBasics of **Geometry**: Introduction, General Terms, Videos and Examples

www.mathopenref.com introduction to Coordinate **Geometry** and the Cartesian Plane - Math Open Reference softschools.comArea of Rectangles and Squares www.mathsisfun.com Interior Angles of Polygons

www.varsitytutors.comDec 11, 2020How to find the area of a rectangle - Basic **Geometry**

softschools.comDec 11, 2020www.varsitytutors.comNov 27, 2020 Area of Rectangles and Squares

www.mathwarehouse.comNov 27, 2020Polygons: Formula for Exterior Angles and Interior Angles, illustrated examples with practice problems on how to calculate..

www.mathsisfun.comNov 26, 2020Interior Angles of Polygons
www.smartick.com Geometric Shapes & Types of Shapes | Smartick

↑http://www.mathsisfun.com/geometry/triangles-congruent-finding.html

Learning Geometry

www.ingramcontent.com/pod-product-compliance
Lightning Source LLC
Chambersburg PA
CBHW070352220526
45467CB00001B/352